AS Biology
UNIT 1

Edexcel

Unit 1: Molecules and Cells

Alan Clamp

To Aristos, Dave, Jon, Jules, Paul and Pete.

Philip Allan Updates
Market Place
Deddington
Oxfordshire
OX15 0SE

Tel: 01869 338652
Fax: 01869 337590
e-mail: sales@philipallan.co.uk
www.philipallan.co.uk

© Philip Allan Updates 2000
Revised September 2002

ISBN 0 86003 469 0

This Guide has been written specifically to support students preparing for the Edexcel AS Biology Unit 1 examination. The content has been neither approved nor endorsed by Edexcel and remains the sole responsibility of the author.

Typeset by Magnet Harlequin, Oxford
Printed by Information Press, Eynsham, Oxford

Contents

Introduction

■ ■ ■

Content Guidance

■ ■ ■

Questions and Answers

Introduction

About this guide

This unit guide is the first book in a series of four, which together cover the whole Edexcel specification for AS and A-level biology. Its aim is to help you prepare for Unit Test 1 in AS Biology which examines the content of **Unit 1: Molecules and Cells**. As this unit is also Unit 1 of AS Biology (Human), this guide is equally relevant to students following this course. There are three sections to this guide:

- **Introduction** — this provides advice on how to use the unit guide, an explanation of the skills required in AS biology and suggestions for effective revision. It concludes with guidance on how to succeed in the unit test.
- **Content Guidance** — this summarises the specification content of Unit 1.
- **Questions and Answers** — this provides two Unit 1 mock test papers for you to try, together with sample answers to these questions and examiner's comments on how these answers could have been improved.

An effective way to use this book is to read through this Introduction section at the beginning of your course to familiarise yourself with the skills required in AS biology. Try to make a habit of using the study skills and revision advice suggested in this section. It may also help to refer back to this information at regular intervals during your course. The Content Guidance section will be useful when you are revising a topic because it highlights the main points of each subsection of the Unit 1 of the specification. You may want to 'tick-off' topics as you learn them to make sure that you have revised everything thoroughly. Finally, the mock tests in the Question and Answer section will provide some very useful practice when preparing for the unit test.

The specification

In order to make a good start to Unit 1, it is important to have a close look at the specification (syllabus). Your teacher should have one, or you can obtain your own copy from the awarding body (Edexcel). In addition to describing the content of the units, the specification provides information about the unit test. It is important for you to understand the key terms used in the specification, as defined below.

- **Recall** — identify and revise biological knowledge gained from previous studies of biology.
- **Know** — be able to state facts, or describe structures and processes, from material within the unit.

- **Understand** — explain the underlying principles and apply this knowledge to new situations.
- **Appreciate** — be aware of the importance of biological information, without having a detailed knowledge of the underlying principles.
- **Discuss** — give a balanced, reasoned and objective review of a particular topic.
- **Describe** — provide an accurate account of the main points (an explanation is not necessary).
- **Explain** — give reasons, with reference to biological theories.

The specification also provides information about the skills required in AS biology. For example, in Unit 1 approximately two-thirds of the marks are available for showing *knowledge and understanding* of biological information, and one-third of the marks are available for *applying* this knowledge and understanding to explain experimental data or solve problems in unfamiliar situations.

Finally, in addition to looking at the specification, it would also be useful for you to read the examiners' reports and published mark schemes from previous unit tests (these are available from Edexcel). These documents show you the depth of knowledge that examiners are looking for in answers, as well as pointing out common mistakes and providing advice on how to achieve good grades in the tests.

Study skills and revision strategies

Students need to develop good study skills if they are to be successful. This section of the Introduction provides advice and guidance on how to study AS biology and suggests some strategies for effective revision.

Organising your notes

Biology students usually accumulate a large quantity of notes and it is useful to keep this information in an organised manner. The presentation of notes is important; good notes should always be clear and concise. You could try organising your notes under headings and subheadings, with key points highlighted using capitals, italics or colour. Numbered lists are useful, as are tables and diagrams. It is a good idea to file your notes in specification order, using a consistent series of informative headings, as illustrated below.

> **UNIT 1 (Molecules and Cells)**
> **Eukaryotic cells: the structure and roles of the major organelles**
> *Mitochondria*
> *Mitochondria are rod-shaped organelles found in the cytoplasm of the cell...*

After the lessons, it is a good idea to check your notes using your textbook(s) and fill in any gaps in the information. Make sure you go back and ask the teacher if you are unsure about anything, especially if you find conflicting information in your class notes and textbook.

Organising your time

When trying to organise your time, it is a good idea to make a revision timetable. This should allow enough time to cover all the material, but also be realistic. For example, it is useful to leave some time at the end of the timetable, just before the unit test, to catch up on time lost, for example through illness. You may not be able to work for very long at a single session — probably no more than 1 hour without a short break of 10–15 minutes. It is also useful to use spare moments, such as when waiting for a bus or train, to do short snippets of revision. These 'odd minutes' can add up to many hours.

Improving your memory

There are several things you can do to improve the effectiveness of your memory for biological information. Organising the material will help, especially if you use topic headings, numbered lists and diagrams. Repeatedly reviewing your notes will also be useful, as will discussing topics with teachers and other students. Finally, using mnemonics (memory aids), such as **A**rteries carry blood **A**way from the heart, can make a big difference.

Revision strategies

To revise a topic effectively you should work carefully through your notes, using a copy of the specification to make sure you have not missed anything out. Summarise your notes to the bare essentials, using the tips given on note-making above. Finally, use the content guidance and mock examinations in this book, discussing any difficulties with your teachers or fellow students.

In many ways, a student should prepare for a unit test like an athlete prepares for a major event, such as the Olympic Games. The athlete will train every day for weeks or months before the event, practising the required skills in order to achieve the best performance on the day. So it is with test preparation: everything you do should contribute to your chances of success in the unit test. The following points summarise some of the strategies that you may wish to use to make sure that your revision is as effective as possible.

- Use a revision timetable.
- Ideally, revise in a quiet room, sitting at a desk or table, with no distractions.
- Test yourself regularly to assess the effectiveness of your revision.
- Practise previous test questions to highlight gaps in your knowledge and understanding and to improve your technique.

- Active revision is much better than simply reading over material. Discuss topics, summarise notes and use the mock tests included in this book to increase the effectiveness of your revision.

The unit test

Unit Test 1 consists of about nine compulsory questions allocated from 4 to 12 marks each, presented in a question–answer booklet. There are 70 marks available in the test and it lasts for 1 hour and 15 minutes (giving you approximately 1 minute per mark). The shorter questions are designed mainly to test knowledge and understanding of the unit content. The longer questions also test skills of interpretation of data that are related to the content of the unit. There is normally one free-prose question on each paper (see p. 8).

There are a number of terms commonly used in unit tests. It is important that you understand the meaning of each of these terms and that you answer the question appropriately.

- **Calculate** — carry out a calculation, showing your working and providing the appropriate units.
- **Compare** — point out similarities *and* differences.
- **Define** — give a statement outlining what is meant by a particular term.
- **Describe** — provide an accurate account of the main points. An explanation is *not* necessary.
- **Discuss** — describe and evaluate, putting forward the various opinions on a topic.
- **Distinguish between** — point out differences only.
- **Explain** — give reasons, with reference to biological facts. A description is *not* required.
- **Outline** — give a brief account.
- **Significance** — the relevance of an idea or observation.
- **State** — give a concise, factual answer (also applies to **give** or **name**).
- **Suggest** — use biological knowledge to put forward an appropriate answer in an unfamiliar situation.
- **What/Why/Where** — these indicate direct questions requiring concise answers.

Whatever the question style, you must read the question *very carefully*, underline key words or phrases, think about your response and allocate time according to the number of marks available. Further advice and guidance on answering test questions is provided in the Question and Answer section at the end of this book.

Structured questions

These are short-answer questions which may require a single-word answer, a short sentence, or a response amounting to several sentences. Answers should be clear, concise and to the point. The marks allocated and the space provided for the answer usually give an indication of the amount of detail required. Typical question styles include:

- naming parts on diagrams
- filling in gaps in a prose passage
- completing tables and tick-boxes
- plotting graphs
- performing calculations
- interpreting experimental data

Free-prose questions

These questions enable you to demonstrate the depth and breadth of your biological knowledge, as well as your ability to communicate scientific ideas in a concise and clear manner. The following points should help you to perform well when answering free-prose questions.

- Make your points clearly and concisely, illustrating with examples where appropriate.
- Try to avoid repetition and keep the answer relevant (refer back to the question).
- The points you make should cover the *full range* of the topics addressed in the question.
- Use diagrams only if appropriate and where they make a useful contribution to the quality of your answer.
- Spend the appropriate amount of time on the question (proportional to the marks available).

The day of the unit test

On the day of the test, make sure that you have:

- two or more blue/black pens, and two or more pencils
- your calculator plus spare batteries
- a watch to check the time
- a ruler and an eraser

Read each question very carefully so that your answers are appropriate. Make sure that you write legibly (you will not be given marks if the examiner cannot read what you have written) and try to spell scientific terms accurately. If you need more room for your answer, look for space at the bottom of the page, the end of the question or after the last question, or use supplementary sheets. If you use these spaces, or sheets, alert the examiner by adding 'continued below' or 'continued on page X'.

Time is often a problem. Make sure that you know how long you have got for the whole test and how many questions you have to do in this time. You could use the number of minutes per mark to work out approximately how long you have for each question (e.g. 7 minutes for a 7-mark question in Unit Test 1).

Do not write out the question, but try to make a number of valid points that correspond to the number of marks available. If you get stuck, make a note of the question number and move on. Later, if you have time, go back and try that difficult question again. Finally, it is a good idea to leave a few minutes at the end to check through the paper, correcting any mistakes or filling in any gaps.

Content
Guidance

Unit 1 of the Edexcel specification includes basic information that you will need for the whole AS or A-level course. A good understanding of this unit will provide you with a solid foundation on which to build your studies of the topics in later units. As its title suggests, this unit is concerned with biological molecules and cells, and it is divided into four topics:

(1) Biological molecules

(2) Enzymes

(3) Cellular organisation

(4) The cell cycle

You may be familiar with some of the information in this unit, but it is important that you know and understand this information exactly as described in the specification. This summary of the specification content will highlight the key points and should prove very useful when learning and revising biology.

Biological molecules

This topic includes the study of water and the major organic molecules found in living organisms: carbohydrates, lipids, proteins and nucleic acids. Before learning the structures and functions of these molecules, it would be useful to revise some background chemistry. The following ten definitions should prove useful when studying this part of the unit.

Atom — the smallest part of an element that cannot be broken down further by chemical means. An atom is made up of a nucleus containing **protons** (particles with a positive charge) and **neutrons** (particles with no charge), surrounded by orbiting **electrons** (particles with a negative charge). Atoms may combine to form molecules.

Condensation reaction — a chemical reaction in which two molecules are joined together and a molecule of water is removed. Condensation reactions are involved in the formation of many biologically important polymers or macromolecules, such as polysaccharides.

Electron — a negatively charged particle normally found orbiting the nucleus of an atom. Electrons are important in making bonds with other atoms to form molecules. If the number of electrons is equal to the number of protons (positively charged particles in the nucleus) then the atom or molecule will have no charge. Otherwise it will be a negatively charged ion (more electrons than protons) or a positively charged ion (more protons than electrons).

Element — any substance that cannot be broken down further by chemical means. Carbon (C), hydrogen (H), oxygen (O) and nitrogen (N) are all elements. Combinations of atoms of the same or different elements, such as carbon dioxide (CO_2), oxygen gas (O_2) or glucose ($C_6H_{12}O_6$), are called molecules.

Hydrogen bond — a weak electrostatic attraction between a hydrogen atom and an oxygen atom. In certain molecules, hydrogen atoms tend to have a slightly positive charge and oxygen atoms tend to have a slightly negative charge, meaning that hydrogen bonding can take place. For example, hydrogen bonds occur between water molecules.

Hydrolysis — the breakdown of a molecule by the addition of water. Many biological molecules are broken down by hydrolysis during digestion. The bonds between the subunits of the large food molecules are broken by adding water. For example, the hydrolysis of a lipid yields fatty acids and glycerol.

Molecule — a substance formed by the combination of atoms. The atoms comprising a molecule may be the same, as in nitrogen gas (N_2), or different, as in carbon dioxide (CO_2). Some biological molecules, such as DNA, contain hundreds or thousands of atoms and are known as macromolecules.

Oxidation — the removal of an electron from a molecule. Oxidation is also used to describe the removal of hydrogen or the addition of oxygen to a molecule. The oxidation of one molecule is always accompanied by the reduction of another.

pH — a measure of the acidity or alkalinity of a solution. pH is inversely related to the concentration of hydrogen ions in a solution. It is measured on a scale of 0 to 14, where 0 indicates a very strong acid, 14 indicates a very strong alkali and 7 is neutral. Any solution or substance that resists changes in pH by taking in or releasing hydrogen ions is known as a buffer.

Reduction — the addition of an electron to a molecule. Reduction is also used to describe the addition of hydrogen or the removal of oxygen from a molecule. The reduction of one molecule is always accompanied by the oxidation of another.

Water

Water is a simple molecule, composed of two hydrogen atoms joined to an oxygen atom (H_2O). Water molecules are slightly positive at one end and slightly negative at the other and are therefore known as **dipolar** molecules. As a result, the molecules have an attraction for each other and form **hydrogen bonds** with neighbouring molecules. The unique chemical and physical properties of water mean that it has a large number of roles in living organisms and in the environment. It is particularly important, for example, as a solvent in living systems.

Carbohydrates

The table below summarises a number of biologically important carbohydrates.

Carbohydrate	Examples	Monosaccaride subunits	Role in living organisms
Monosaccharides	Triose	N/A	Important intermediate in metabolism
	Glucose	N/A	Energy source in respiration
	Ribose	N/A	Part of RNA
Disaccharides	Sucrose	Glucose + fructose	Transported in the phloem of plants
	Lactose	Glucose + galactose	Main carbohydrate in milk
	Maltose	Glucose	Found in germinating seeds
Polysaccharides	Glycogen	α-Glucose	Storage carbohydrate in animals
	Starch	α-Glucose	Storage carbohydrate in plants
	Cellulose	β-Glucose	Part of plant cell walls
N/A = not applicable			

Some important carbohydrates

Carbohydrates are compounds that contain the elements carbon, hydrogen and oxygen, with the general formula $C_x(H_2O)_y$. The simplest carbohydrates are **monosaccharides**, which usually contain three carbon atoms (trioses), five carbon atoms (pentoses) or six carbon atoms (hexoses). **Disaccharides** are formed from a condensation reaction (a chemical reaction in which two molecules are joined together and one molecule of water is removed) between two hexose monosaccharides. **Polysaccharides** are made up of many hexose monosaccharides linked by condensation reactions.

Lipids

Lipids are a large group of organic substances that are insoluble in water but soluble in organic solvents such as ethanol. A number of important biological molecules are classified as lipids, including the following.

- **Triglycerides** — composed of fatty acids and glycerol. They function as energy stores in plants and animals.
- **Phospholipids** — composed of fatty acids, glycerol and a phosphate group. They are important components of cell membranes.
- **Steroids** — ring-based structures that do not contain fatty acids. Steroids include a number of hormones, such as oestrogen and testosterone.

Proteins

Proteins are a large group of organic molecules composed of long chains of amino acids. The amino acids in a protein are joined together by peptide bonds. The long chains of amino acids (known as polypeptides) may have four different levels of structure.

- **Primary structure** — the sequence of amino acids.
- **Secondary structure** — the coiling of a polypeptide into a helix.
- **Tertiary structure** — the irregular folding of a polypeptide into a globular shape.
- **Quaternary structure** — the association of more than one polypeptide chain.

Many of the functions of proteins can be explained in terms of their structure. Fibrous proteins are usually long, coiled strands or flat sheets and have a structural function. They include collagen, actin, myosin and fibrin. Globular proteins have a roughly spherical shape and have a physiological function. They include enzymes, antibodies, haemoglobin and insulin.

Nucleic acids

Nucleic acids, such as ribonucleic acid (**RNA**) and deoxyribonucleic acid (**DNA**), are biological molecules consisting of long chains of **nucleotides**. A nucleotide consists of a pentose (5-carbon) sugar, a nitrogenous base and a phosphate group. The pentose sugar may be either ribose or deoxyribose, and the nitrogenous base may be adenine, cytosine, guanine, thymine or uracil. Nucleotides may join together by condensation reactions to form polynucleotides, such as DNA or RNA.

DNA is a molecule that forms the genetic material of all living organisms. It consists of two polynucleotide chains coiled into a double helix. A simplified diagram of part of a DNA molecule is shown below.

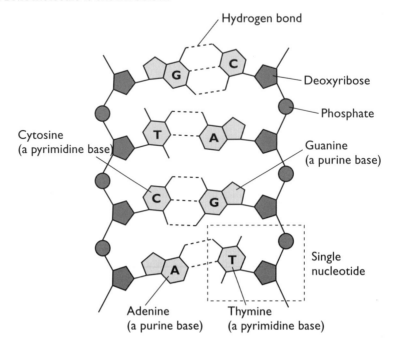

Note: Adenine (A) always bonds with thymine (T)
Guanine (G) always bonds with cytosine (C)

The molecular structure of DNA

DNA is a major constituent of the chromosomes in the nucleus of a eukaryotic cell. It is also found in a non-chromosomal form in prokaryotic cells, chloroplasts, mitochondria and viruses. DNA plays a central role in the determination of hereditary characteristics by controlling protein synthesis in cells. A section of DNA that codes for the production of a particular protein or polypeptide is known as a **gene**.

DNA replication is the process by which a DNA molecule produces two exact copies of itself. The process is controlled by the enzyme **DNA polymerase**. Hydrogen bonds break between the two polynucleotide chains and the parent DNA molecule unwinds. Each DNA strand then acts as a template for the synthesis of a new strand. Free nucleotides line up opposite their complementary bases and are joined together by condensation reactions to form two new DNA molecules. This process is known as semiconservative replication, as each new molecule contains half the original parent DNA molecule. It is a common mistake to think that DNA replication is the same as transcription, which takes place during protein synthesis (see p.17). There are several important differences between these processes and it is important that you do not get them confused.

```
 ┌T A┐
 ├A T┤
 ├T A┤
 ├C G┤
 ├G C┤
 └A   T┘
```

Semi-conservative replication

The polynucleotide strands of
DNA separate

```
┌T A┐   ┌T A┐
├A T┤   ├A T┤
├T A┤   ├T A┤
├C G┤   ├C G┤
├G C┤   ├G C┤
└A  T┘   └A  T┘
```

Each strand acts as a template
for the formation of a new
molecule of DNA

Individual nucleotides line up
with complementary bases on
parent DNA strands

```
┌T A┐   ┌T A┐
├A T┤   ├A T┤
├T A┤   ├T A┤
├C G┤   ├C G┤
├G C┤   ├G C┤
└A T┘   └A T┘
```

The nucleotides are joined
together by DNA polymerase
to form two molecules of DNA

Each molecule contains a strand
from the parent DNA and a
new strand

There are three forms of RNA: messenger RNA (**mRNA**), transfer RNA (**tRNA**) and ribosomal RNA (**rRNA**). The structure of RNA is similar to that of DNA, in that they are both polynucleotides. However, RNA contains the sugar **ribose** rather than deoxyribose and it does not contain the organic base thymine (T), but contains **uracil** (U) instead.

DNA and RNA are both involved in **protein synthesis**, which is the synthesis of a protein from its constituent amino acids according to the sequence of bases in DNA that constitute a gene. The process of protein synthesis takes place on **ribosomes** in the cell and is summarised in the diagram below.

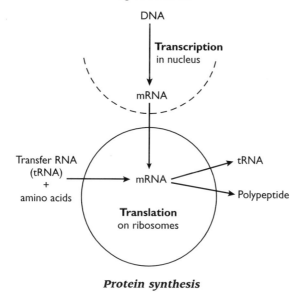

Protein synthesis

Biochemical tests

You are expected to know how to carry out biochemical tests for starch, reducing and non-reducing sugars, and proteins. A summary of these tests is provided in the table below.

Test	Procedure
Biuret test for proteins	Add sodium hydroxide to the test sample, followed by a few drops of copper sulphate solution. If protein is present the solution turns purple.
Benedict's test for reducing sugars (e.g. glucose)	Heat the test sample with Benedict's reagent. If a reducing sugar is present an orange-red precipitate is formed.
The test for non-reducing sugars (e.g. sucrose)	First heat the test sample with Benedict's reagent to confirm that there is no reducing sugar present. Then hydrolyse the sugar by heating with dilute acid followed by neutralising with sodium hydrogen carbonate. Finally repeat Benedict's test. If a non-reducing sugar is present, this second test will give an orange-red precipitate.
Iodine test for starch	Add iodine solution to the test sample. If starch is present the solution turns blue-black.

Biochemical tests

Enzymes

Enzymes are protein molecules that act as catalysts in living organisms. They are globular proteins that speed up biochemical reactions by lowering the activation energy (the energy required to activate or begin a chemical reaction) required for a particular reaction.

The individual reactions that constitute metabolism are catalysed by specific enzymes, which act according to the lock-and-key hypothesis or the induced-fit hypothesis. Essentially, enzymes combine with a specific substrate and convert it to a product, according to the equation below.

$$E + S \rightleftharpoons ES \rightleftharpoons EP \rightleftharpoons E + P$$

| Enzyme + substrate | Enzyme–substrate complex | Enzyme–product complex | Enzyme + product |

The lock-and-key hypothesis proposes that the active site of an enzyme has a very specific shape (like a lock) into which the substrate molecule (the key) fits exactly, to form an enzyme–substrate complex. This model is generally considered to be less useful than the induced-fit hypothesis, which argues that the interaction of an enzyme with its substrate causes the active site to change shape slightly in order to allow the formation of an enzyme–substrate complex.

Several factors influence the rate of enzyme-controlled reactions. These include the concentration of enzyme and substrate, pH and temperature. Large changes in pH, or high temperatures, will denature the enzyme, altering the shape of the active site. If this happens, the substrate will no longer fit the site and the enzyme will cease to function.

Unit test questions will often ask you to explain the influence of a certain factor on the rate of an enzyme-controlled reaction. Make sure that you do not simply *describe* the effect if an *explanation* is needed. Remember that excessive heat denatures enzymes by disrupting their tertiary structure, it does not kill them! Low temperatures do not denature enzymes, they simply work more slowly because there is less energy in the system.

Enzymes are also affected by the presence of coenzymes, activators or inhibitors. There are two main forms of enzyme inhibition.

Active site-directed inhibition — a reduction in the rate of an enzyme-catalysed reaction by a molecule that is similar in shape to the normal substrate. Inhibitor molecules compete with substrate molecules for the active site of an enzyme and the extent of the inhibition depends upon the relative concentrations of inhibitor and substrate.

Non-active site-directed inhibition — the inhibition of enzyme activity due to the binding of an inhibitor molecule at a site other than the active site. The binding of the inhibitor changes the shape of the active site, preventing the enzyme from functioning and reducing the rate of reaction. Unlike active site-directed inhibition, this type of inhibition cannot be reversed by increasing the concentration of substrate.

Many enzymes have important commercial applications, such as the use of pectinases in the production of fruit juice and proteases in biological detergents. It is often an advantage to use enzyme immobilisation in a commercial context. Immobilisation is a technique used in biotechnology in which enzymes are fixed to unreactive materials, such as beads, which are then placed in a column. The substrate is converted into products as it passes down the column. There are several advantages to immobilisation. The process is cheaper because the enzyme can be easily recovered at the end of the reaction and used again. It also helps avoid contamination of the products and can make certain enzymes tolerant of higher temperatures and a wider pH range. Some biology students seem to think that the immobilisation of an enzyme stops it from working — make sure you do not make that mistake!

Finally, you will be expected to have undertaken practical work to investigate:

- the effects of temperature, pH and enzyme concentration on enzyme activity
- enzyme immobilisation using lactase
- the use of pectinase in the production of fruit juice

Cellular organisation

Cells

Cells are the basic structural and functional units of most living organisms. A cell comprises a mass of jelly-like **cytoplasm** which contains a number of **organelles** and is surrounded by a **cell membrane**. There are two basic types of cell.

- **Prokaryotic** cells (e.g. bacteria) have no nucleus and very few organelles
- **Eukaryotic** cells (e.g. human liver cells and palisade mesophyll cells in the leaves of green plants) have a nucleus and a large number of different organelles

A common unit test question asks you to state the differences between prokaryotic and eukaryotic cells, as summarised in the table below.

Feature	Prokaryotic cells	Eukaryotic cells
Size	Cells smaller, usually less than 5 µm in diameter.	Cells larger, often as much as 50 µm in diameter.
Nucleus and DNA	Cells do not have a nucleus. The DNA, which is not associated with proteins, is present as a circular strand in the cytoplasm of the cell.	Cells have a nucleus. The DNA, which is in long strands, is associated with proteins, forming chromosomes.
Organelles	Few organelles present and none of them is surrounded by a membrane.	Many membrane-surrounded organelles, such as mitochondria, are present.
Ribosomes	Only have small ribosomes which are free in the cytoplasm.	Have small ribosomes and larger ones which are associated with membranes, forming rough endoplasmic reticulum.

Comparison of prokaryotic and eukaryotic cells

The Edexcel specification requires you to understand the structure and functions of a number of organelles found in cells, as listed below.

Cell membrane — a membrane that encloses a cell (the cell surface membrane) or which surrounds certain organelles within a cell. Cell membranes are typically about 7 nm thick and consist mainly of phospholipids and proteins, arranged as a fluid-mosaic. The membranes are partially permeable, regulating the movement of substances in and out of the cell or organelles.

Cell wall — a rigid structure found outside the cell surface membrane of plant and bacterial cells. The main component of cell walls in plants is **cellulose** and in bacteria it is **peptidoglycan** (a complex polymer of polysaccharides and amino acids). Cell walls are very strong and provide support for the cell and for the whole organism.

Centriole — a structure found in animal cells which plays a role in **spindle formation** during cell division. Centrioles are usually found in pairs and they are small, hollow cylinders, each containing a ring of microtubules.

Chloroplast — an organelle found in the cells of plants which is the site of **photo-synthesis**. Chloroplasts are small, flattened discs with a double outer membrane.

Endoplasmic reticulum — a system of membranes found in the cytoplasm of eukaryotic cells. There are two types of endoplasmic reticulum (ER) in cells:

- **rough ER** has an outer surface lined with ribosomes and is involved in the synthesis of proteins
- **smooth ER** has no ribosomes and is concerned with the synthesis of lipids and steroids

Flagellum — a long hair-like structure found on the surface of cells which is involved in locomotion. Flagella (plural) are up to 150 µm long and are found on the surface of bacteria and sperm cells. The possession of a flagellum makes these cells motile.

Golgi apparatus — a collection of membranes and vesicles found in the cytoplasm of cells. The Golgi apparatus is an organelle with a number of functions, including the synthesis of glycoproteins, the secretion of enzymes and hormones and the formation of lysosomes.

Lysosome — a small membrane-bound organelle found in the cytoplasm of many eukaryotic cells. Lysosomes are formed from the Golgi apparatus and contain **hydrolytic enzymes**. They are used by certain cells in phagocytosis, to break down the ingested material.

Mitochondrion — a rod-shaped organelle found in the cytoplasm of eukaryotic cells. The function of mitochondria is to carry out the chemical reactions involved in **aerobic respiration**.

Nucleus — a large organelle found in eukaryotic cells. The nucleus contains **DNA** in the form of **chromosomes** and is the site of synthesis of RNA. It is separated from the rest of the cell by a double membrane (envelope), which has pores to allow the movement of substances in or out.

Plasmid — a small circular molecule of DNA found in bacteria. Plasmids sometimes contain important genes, such as those conferring resistance to antibiotics.

Ribosome — a small organelle which acts as the site of **protein synthesis** in both prokaryotic and eukaryotic cells. Ribosomes are spherical structures composed of ribosomal RNA and protein. They are found free in the cytoplasm and attached to endoplasmic reticulum (rough ER).

Finally, you will be expected to have carried out the following practical work in relation to your work on cellular organisation:

- setting up and using a light microscope
- making accurate drawings of cells and plans of tissues
- using a graticule to make measurements and to calculate scale

Transport across membranes

Molecules and ions move in and out of cells in four ways, as summarised in the table below.

Mechanism	Substances transported	Requires energy in the form of ATP	Takes place against a concentration gradient	Requires protein carrier molecules
Diffusion	Various	No	No	No
Facilitated diffusion	Various	No	No	Yes
Osmosis	Water only	No	No	No
Active transport	Various	Yes	Yes	Yes

Mechanisms of transport across cell membranes

In addition to the mechanisms listed above, substances move in and out of cells in two other ways.

Endocytosis — the uptake of large particles or fluids through the surface membrane of a cell. During endocytosis, the cell surface membrane surrounds the particle and forms a vesicle or vacuole, which moves into the cytoplasm.

Exocytosis — the transport of large particles or fluids out of a cell via the cell surface membrane. Vesicles are formed from the Golgi apparatus in the cytoplasm of the cell. These fuse with the surface membrane, releasing their contents outside the cell.

It is easy to confuse these two terms. Remember that *endo* means 'into' or 'within', and *exo* means 'out of' or 'outside' (it may help you to remember to think of the words **en**trance and **ex**it).

Aggregations of cells

The body of an individual organism contains many different types of cell, each type being specialised and adapted for a particular function. A group of similar cells that carry out a given function is known as a tissue (e.g. muscle tissue). Different tissues make up organs (e.g. the stomach) and several organs can combine to form a system (e.g. the digestive system).

The cell cycle

Chromosome structure

Chromosomes are thread-like structures, composed of DNA and protein, found in the nucleus of plant and animal cells. During the early stages of cell division, chromosomes consist of a pair of **chromatids** held together by a **centromere**. Chromosomes are not visible in non-dividing cells, when they assume an elongated form of DNA wrapped around proteins (histones), known as **chromatin**.

Each chromosome carries a number of **genes**. The number of chromosomes in each cell is constant for a particular species. For example, human body cells have the diploid number of 46 chromosomes (or 23 homologous pairs). Gametes, however, contain only half this number (the haploid number), which is 23 in humans.

The cell cycle

The cell cycle is the sequence of events that occurs during cell growth and cell division. In order to understand the cell cycle, it is necessary to have a good background knowledge of cell structure and an understanding of the replication of DNA. The cell cycle can be divided into four main stages, as shown in the figure below.

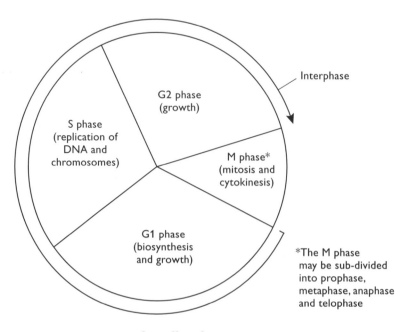

The cell cycle

Mitosis

Mitosis is a type of cell division used for growth, repair and the replacement of cells. It involves the production of genetically identical daughter cells (clones), which have the same chromosome number (the diploid number) as the parent cell. The process of mitosis is summarised in the following table.

Stage	Key events
Prophase	Chromosomes shorten and thicken. The nuclear membrane starts to break down and the spindle forms.
Metaphase	Chromosomes line up on the equator of the cell, attached to spindle fibres by their centromeres.
Anaphase	Chromatids separate and move to opposite ends of the dividing cell. Each chromatid has now become a chromosome.
Telophase	The spindle fibres break down, the nuclear membrane re-forms, chromosomes elongate and cell division is complete.

The key stages of mitosis

In addition to knowing these descriptions, you are expected to be able to recognise the various stages in diagrams or photographs. Finally, you will be expected to have carried out practical work on the preparation and staining of root tip squashes to study stages in mitosis using a light microscope.

Questions
&
Answers

In this section of the guide there are two mock papers written in the same format as the real unit test papers. All questions are based on the topic areas outlined in the Content Guidance section. When you have completed a paper, ideally under timed conditions (allowing 1 hour 15 minutes per paper — see pp. 7–8), compare your answers with those of Candidate A and Candidate B. Try to avoid looking at the sample answers and examiner's comments before completing the tests. Make sure that you correct any mistakes and that you study the examiner's comments very carefully. You will get a much better grade if you can avoid the common errors made by many candidates in their unit tests.

Examiner's comments

Candidate responses include examiner's comments after each section of the answer. These examiner's comments are preceded by the icon *e* and indicate where credit is due. In the weaker answers, they also point out areas for improvement, specific problems and common errors, such as poor time management, lack of clarity, weak or non-existent development, irrelevance, misinterpretation of the question and mistaken meanings of terms.

Molecules and cells (I)

(1) Read the following passage about water and write on the dotted lines the most appropriate word or words to complete the passage.

In a water molecule, two hydrogen atoms are joined to an .... atom by covalent bonds. When covalent bonds are formed, the .... are not always shared equally, causing a water molecule to be slightly positive at one end and slightly negative at the other. Water is therefore known as a .... molecule. Water molecules have an attraction for each other and form .... bonds with neighbouring molecules.

4 marks

(2) The statements in the table below refer to three polysaccharide molecules. If the statement is correct, place a tick (✔) in the appropriate box and if the statement is incorrect, place a cross (✘) in the appropriate box.

Statement	Glycogen	Starch	Cellulose
Energy store in plant cells			
Glycosidic bonds present			
Polymer of α-glucose ·			
Unbranched chains only			

4 marks

(3) The table below describes three organelles commonly found in eukaryotic cells. Complete the table by writing the name of the organelle, its description or its main function in the boxes provided.

Name of organelle	Description	Function
Cell surface membrane		
	Small spherical organelles, found attached to endoplasmic reticulum or free in the cytoplasm	Protein synthesis
	Rod-shaped structures with a double membrane, the inner one folded to form cristae	

5 marks

(4) The diagram below shows the structure of a phospholipid molecule.

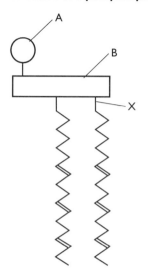

(a) Name the parts labelled **A** and **B**. (2 marks)

(b) Name the chemical bond labelled **X** on the diagram. (1 mark)

(c) Is this phospholipid saturated or unsaturated? **Give a reason for your answer.** (1 mark)

(d) Phospholipids are a major component of cell surface membranes. Name **two**
other types of biological molecule found in cell surface membranes. (2 marks)

Total: 6 marks

(5) The diagram below shows the molecular structure of deoxyribonucleic acid (DNA).

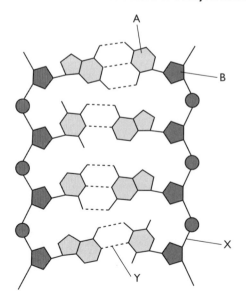

(a) (i) Name the parts labelled **A** and **B**.   (2 marks)

   (ii) Name the types of bond labelled **X** and **Y** on the diagram.   (2 marks)

(b) **DNA** from a daffodil plant was analysed and the number of molecules of each type of organic base was found. The results of this analysis are shown in the table below.

Organic base	Number of molecules
Adenine	32 400
Cytosine	46 600
Guanine	45 300
Thymine	34 900

  (i) Which two bases in the table are purines?   (1 mark)

  (ii) What do the results of the analysis suggest about base pairing in the **DNA** of daffodils?   (3 marks)

Total: 8 marks

(6) (a) The table below refers to features which may relate to facilitated diffusion, osmosis and active transport. If the feature applies, place a tick (✔) in the appropriate box and if the feature does not apply, place a cross (✖) in the appropriate box.

Feature	Facilitated diffusion	Osmosis	Active transport
Requires energy in the form of ATP			
Can take place against a concentration gradient			
Requires protein carrier molecules			

(3 marks)

(b) The rate of diffusion through a membrane is proportional to:

$$\frac{\text{surface area} \times \text{difference in concentration}}{\text{thickness of membrane}}$$

Predict whether the values of each of the three variables will be high or low when the rate of diffusion through the membrane is at a maximum.

          Surface area   ....

          Difference in concentration   ....

          Thickness of membrane   ....

(1 mark)

(c) The diagram below shows three adjacent plant cells.

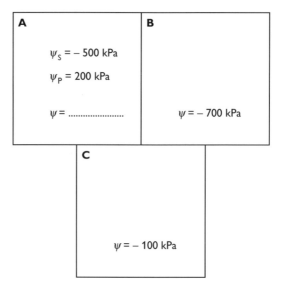

(i) Calculate the water potential of cell **A**. Write your answer in the space provided on the diagram. (1 mark)

(ii) Show, by means of arrows on the diagram, the direction of water movement between these cells. (2 marks)

The graph below shows how the rate of uptake of glucose by muscle cells varies with the glucose concentration of the surrounding tissue fluid.

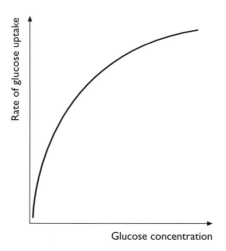

(d) Explain how these data support the hypothesis that the cells take up glucose by facilitated diffusion. (2 marks)

Total: 9 marks

(7) Catalase is an enzyme that breaks down hydrogen peroxide into oxygen and water. An experiment was carried out to investigate the effect of temperature on the activity of catalase and the results are shown in the graph below.

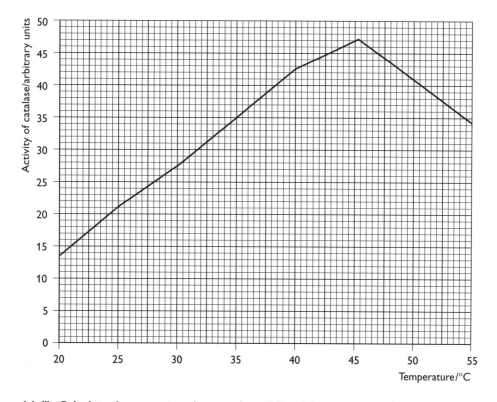

(a) (i) Calculate the percentage increase in activity of the enzyme as the temperature increases from 35 °C to 45 °C. Show your working. (3 marks)

(ii) Explain why temperature affects enzyme activity as shown in the graph. (4 marks)

(b) Describe how this experiment could be modified to investigate the effect of pH on the activity of catalase. (5 marks)

Total: 12 marks

(8) The diagram below shows the sequence of bases in one strand of DNA from part of a gene. The base sequence is read from left to right.

G	G	T	C	A	C	T	T	C	G	A	G	C	C	C

(a) Explain what is meant by the term *gene*. (2 marks)

(b) Write out the sequence of bases in mRNA that would be coded for by this DNA base sequence. (2 marks)

The table below shows the anticodons of some tRNA molecules and the specific amino acids each would carry.

tRNA anticodon	Amino acid
AGC	Serine
CAC	Valine
UUC	Lysine
GGU	Proline
GAG	Leucine
CCC	Glycine
CGC	Alanine

(c) Using the information in the table, write down the amino acid sequence coded for by this part of the gene. (2 marks)

The diagram below shows the same length of DNA after it has undergone a mutation (a change in the sequence of bases).

| G | G | T | C | G | C | T | T | C | G | A | G | C | C | C |

(d) (i) Describe the effect of this mutation. (2 marks)
(ii) Suggest how this mutation might affect the protein produced. (4 marks)

Total: 12 marks

(9) Give an account of the process of mitosis. **10 marks**

ock paper

Molecules and cells (II)

(1) The table below refers to features of prokaryotic and eukaryotic cells. If the feature is present, place a tick (✔) in the appropriate box and if the feature is absent, place a cross (✘) in the appropriate box.

Feature	Prokaryotic cell	Eukaryotic cell
Ribosomes		
Endoplasmic reticulum		
Mesosomes		
Golgi apparatus		

4 marks

(2) The table below summarises some of the biochemical tests for proteins and carbohydrates. Complete the table by writing the name of the substance being tested, the name of the test, or a description of a positive result in the boxes provided.

Substance	Name of test	Positive result
		Biuret test
Reducing sugars		Orange-red precipitate
		Iodine test

5 marks

(3) Read the following passage about lipids and write on the dotted lines the most appropriate word or words to complete the passage.

Lipids are fats and oils formed by .... reactions between .... and fatty acids. During the reaction, a triglyceride is formed and three molecules of .... are removed. The main function of lipids is as an energy store, but they also play important roles in .... and ....

5 marks

(4) Distinguish between each of the following pairs of terms.
  (a) Globular proteins and fibrous proteins.   (3 marks)
  (b) Rough endoplasmic reticulum and smooth endoplasmic reticulum.   (3 marks)

Total: 6 marks

(5) (a) **Explain why root tips are particularly suitable material to use for preparing slides to show mitosis.** (1 mark)

(b) **Give *one* reason for each of the following steps in preparing a slide to show mitosis in cells from a root tip.**

(i) **The tissue should be stained.** (1 mark)

(ii) **The stained material should be pulled apart with a needle and pressure applied to the cover slip during mounting.** (1 mark)

(c) **The graph below shows the movements of chromosomes during mitosis.**

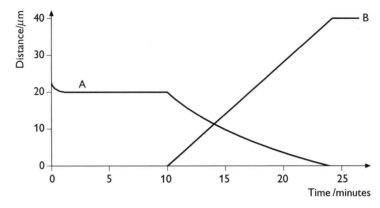

(i) **Curve A shows the mean distance between the centromeres of the chromosomes and the corresponding pole of the spindle. Suggest what line B represents.** (1 mark)

(ii) **At what time did anaphase begin?** (1 mark)

(iii) **Explain how the graph supports your answer to (ii).** (2 marks)

Total: 7 marks

(6) **The graph below shows the energy changes which take place during a chemical reaction.**

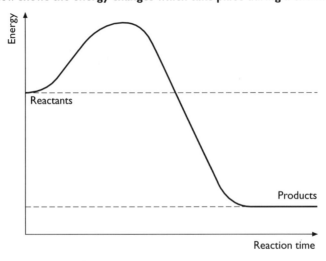

(a) (i) Use the graph to explain what is meant by the term *activation energy*.   (2 marks)
    (ii) Draw a curve on the graph to show the energy changes which would
        take place if the same chemical reaction were catalysed by an enzyme.   (2 marks)

The graph below shows the effect of changing substrate concentration on the rate of an enzyme-controlled reaction.

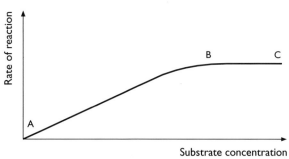

(b) Describe and explain the effect of changes in substrate concentration on the rate of
    reaction.
    (i) Between **A** and **B**   (3 marks)
    (ii) Between **B** and **C**   (3 marks)

Total: 10 marks

(7) The diagram below summarises the flow of information during the synthesis of a
    polypeptide.

                      Stage A   Stage B
        DNA ————————→ mRNA ————————→ polypeptide

(a) (i) Name the stages labelled **A** and **B** on the diagram.   (2 marks)
    (ii) Name the enzyme which catalyses the production of mRNA from DNA.   (1 mark)
    (iii)Where in a cell does stage **B** take place?   (1 mark)
    (iv)Explain how stage **A** differs from the replication of DNA.   (3 marks)
(b) Biochemical analysis of a sample of elephant DNA showed that 33% of the
    organic bases were guanine. Calculate the percentage of bases in the sample
    that would be adenine. Show your working.   (3 marks)
(c) State *two* differences between the structure of DNA and that of RNA.   (2 marks)

Total: 12 marks

(8) (a) Explain what is meant by the term *osmosis*.   (2 marks)
    (b) A plant cell has a solute potential (ψ_S) of –600 kPa and a pressure potential
        (ψ_P) of 200 kPa.
        (i) Calculate the water potential (ψ) of this cell. Show your working.   (2 marks)
        (ii) If this cell was placed in a solution with a water potential (ψ) of –100 kPa,
            would water move in or out of the cell? Explain your answer.   (1 mark)

(c) State **two** ways in which active transport differs from simple diffusion. (2 marks)

(d) The two graphs below show how the rate of entry of molecules into a cell varies with the external concentration of the molecules. Graph **A** is a case of simple diffusion and graph **B** refers to facilitated diffusion.

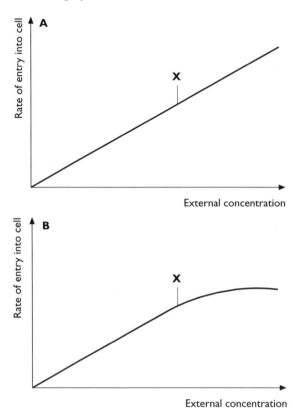

(i) Explain why the shapes of the curves are similar up to point **X**. (2 marks)

(ii) Explain why the shapes of the curves are different after point **X**. (2 marks)

Total: 11 marks

(9) Give an account of the biological significance of polysaccharides. **10 marks**

Answers to mock paper 1: Candidate A

(1) In a water molecule, two hydrogen atoms are joined to an .....oxygen ✓.... atom by covalent bonds. When covalent bonds are formed, the .....electrons ✓.... are not always shared equally, causing a water molecule to be slightly positive at one end and slightly negative at the other. Water is therefore known as a .....dipolar ✓.... molecule. Water molecules have an attraction for each other and form ....ionic ✗..... bonds with neighbouring molecules.

🖉 This answer receives 3 out of 4 marks. The only mistake is the last missing word. Water is *not* ionic and it forms **hydrogen** bonds with other water molecules. Note that for the third answer, either dipolar or **polar** would receive a mark.

(2)

Statement	Glycogen	Starch	Cellulose
Energy store in plant cells	✗	✔	✔ ✗
Glycosidic bonds present	✔	✔	✔
Polymer of α-glucose	✔	✔	✗
Unbranched chains only	✗	✗	✔

🖉 This answer receives 3 out of 4 marks. Although cellulose is found in plant cells, it is *not* an energy store, i.e. it is not broken down to release energy. Cellulose is a polymer of β-glucose which has a structural role in the cell walls of plants. Note that *all* polysaccharides contain glycosidic bonds.

(3)

Name of organelle	Description	Function
Cell surface membrane	Layers of cellulose ✗	Structural support in plant tissues ✗
Ribosomes ✓	Small spherical organelles, found attached to endoplasmic reticulum or free in the cytoplasm	Protein synthesis
Mitochondria ✓	Rod-shaped structures with a double membrane, the inner one folded to form cristae	Respiration ✗

🖉 This answer receives 2 out of 5 marks. Candidate A has clearly confused the cell surface membrane with the *cell wall* (or simply misread the question) and therefore these answers are incorrect. A suitable description for the cell surface membrane could be a **phospholipid bilayer containing proteins** and its function is **to control the movement of substances into and out of the cell**. The answer 'respiration' in the third line is correct, but not specific enough. Mitochondria are responsible for **aerobic** respiration.

(4) (a) A = phosphate ✓   B = fatty acid ✗

Part B is **glycerol**. Remember that phospholipids have a phosphate 'head' and a fatty acid 'tail'. Glycerol is the middle part of the molecule.

(b) Covalent ✗

Although X *is* a covalent bond, this answer is not specific enough. Peptide bonds in proteins and glycosidic bonds in carbohydrates are also covalent, as are many other bonds. The answer required here, which is specific to lipids, is an **ester bond**.

(c) Saturated — because there are double bonds in the fatty acid chains ✗

In this answer the *reason* is correct, but having no double bonds makes the phospholipid **unsaturated**. Note that there is only one mark available. You do not get a mark for simply guessing that the molecule is saturated or unsaturated. A valid reason is needed to support your answer.

(d) protein ✓   and cholesterol ✓

Two good answers at the end give Candidate A 3 out of 6 marks for this question. Note that **carbohydrate, glycoproteins or glycolipids** would also get a mark in part (d).

(5) (a) (i)   A = glucose ✗   B = pentose sugar ✓

Part A is an organic **base**, which could be adenine, thymine, cytosine or guanine. It could not be uracil as this is found only in RNA. Candidate A might have thought it was glucose because it appears to be a six-sided molecule, but glucose is *not* present in DNA.

(ii) X = glycosidic bond ✗   Y = hydrogen bond ✓

X represents the covalent bond joining individual nucleotides together in DNA. It is therefore a **phosphodiester bond**. Glycosidic bonds are found in carbohydrates and *not* in nucleic acids.

(b) (i) cytosine and thymine ✗

Although cytosine and thymine are the same type of molecule, they are both pyrimidines and not purines. The correct answer is **adenine** and **guanine**.

(ii) The ratio of A:T is approximately 1:1; ✓ the ratio of C:G is approximately 1:1 ✓. This suggests that A pairs with T and that C pairs with G ✓.

A very good final answer gives Candidate A 5 out of 8 marks for this question.

(6) (a)

Feature	Facilitated diffusion	Osmosis	Active transport
Requires energy in the form of ATP	✔✗	✗	✔
Can take place against a concentration gradient	✔✗	✗	✔
Requires protein carrier molecules	✔	✗	✔

🖉 This answer receives 1 out of 3 marks. Candidate A appears to have confused facilitated diffusion with active transport. Facilitated diffusion *does* require protein carrier molecules, but it *cannot* take place against a concentration gradient and it does *not* require energy in the form of ATP.

(b)
Surface area .....high...............
Difference in concentration .....high...............
Thickness of membrane .....low ✓............

(c)

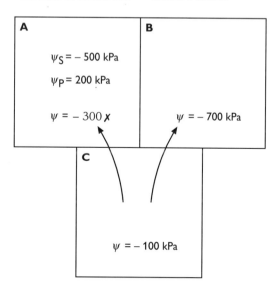

🖉 Although the calculation has been done correctly, i.e. $\psi = \psi_S + \psi_P$, Candidate A has forgotten to put the *units* in the answer. The correct response is −300 **kPa**. The arrows are correct (water moves from a *less negative* to a *more negative* water potential), but Candidate A has not drawn arrows to indicate that water will also move **from A to B** for the second mark. Remember that you usually have to make two separate points for 2 marks. Note that water potential is *never* positive. The highest possible value of ψ is zero, as found in pure water. During osmosis, water moves from a *high* water potential to a *low* water potential.

(d) The uptake increases as the concentration of glucose in the surrounding fluid increases ✓; but uptake then flattens out at high concentrations of glucose as the protein carrier molecules become saturated ✓.

This answer is worth 2 marks, giving Candidate A 5 out of 9 marks for the whole question.

(7) (a) (i) $47 - 35 = 12$ ✓; $(12/100) \times 35 = 4.2\%$ ✗

Candidate A correctly found the figures on the graph corresponding to 35 °C (35 units) and 45 °C (47 units) and realised that it was the *difference* that was important. However, the percentage calculation should have been $(12/35) \times 100 = 34.3\%$. Remember that to calculate a percentage increase, you should use the following formula: (increase/original) $\times$ 100.

(ii) As the temperature increases from 20 °C to 45 °C, the activity of catalase increases from 13.5 units to 47 units (an increase of 33.5 units). Between 45 °C and 55 °C, the activity of catalase falls by 13 units to 34 units.

This would be a very good answer if the question had asked the candidate to *describe* the relationship between temperature and enzyme activity. However, the key word was *explain*, meaning that biological information is required to account for the shape of the graph. As a result, Candidate A scores zero for this part of the question.

(b) Repeat the experiment, but keep temperature constant ✓ and use a range of buffers at different pH ✓; this will show you the effect of pH on the activity of catalase.

This answer is worth 2 out of 4 marks. It is correct, but very brief and it therefore lacks the detail required for a higher mark. For example, specific examples of pH buffers could have been stated, such as **pH3, pH5, pH7, pH9** and **pH11**. Keeping the temperature constant is one important control, but so is keeping the **concentration of substrate and enzyme constant**. Finally, in order to assess accurately the effect of pH on the activity of catalase, **the measurements at each pH should be repeated** and a **graph plotted of enzyme activity against pH**. Candidate A scores 3 out of 12 for this question.

(8) (a) a section of DNA ✓

Only one point has been made in this answer and therefore it is only worth 1 mark. For the second marking point, the function of a gene should be mentioned, i.e. **it codes for the production of a protein**.

(b) CCAGTGAAGCTCGGG ✗

Candidate A has correctly remembered the rules for base-pairing, but has made the common mistake of forgetting that RNA does *not* contain thymine (T) but has uracil (U) instead. This occurs twice and therefore both marks are lost.

(c) proline – valine – lysine – leucine – glycine ✓✓

(d) (i) Adenine changed to guanine ✓.

🔁 Again this is correct, but not sufficient for 2 marks. Candidate A could have gone on to explain *where* the mutation occurred (**fifth base from the left**), or what *effect* it would have on the amino acid sequence (**alanine would be present instead of valine**).

(ii) The amino acid sequence will be changed ✓ so the protein might not function ✓.

🔁 Another brief answer, worth 2 out of 4 marks. Candidate A could have explained that changing the amino acid would **change the primary structure of the protein** and that this may **alter the tertiary structure and therefore shape of the protein**. These extra details would have given the candidate full marks. Overall, Candidate A scores 6 out of 12 marks for the whole question.

(9) Mitosis is a type of cell division concerned with growth. A cell divides to produce two new cells by a series of stages: prophase, anaphase, metaphase and telophase. In prophase, the nuclear membrane breaks down and the chromosomes form homologous pairs. The next stage is anaphase, where the chromosomes attach to spindles inside the cell and line up along the equator. During metaphase one homologous chromosome from each pair moves to the poles of the cell. Telophase is the last stage, where the nuclear membrane re-forms and the new cells have been produced.

🔁 Candidate A correctly identifies that mitosis is important in growth and knows the names of the stages of nuclear division (although not the correct order). The nuclear membrane does break down during prophase, but the chromosomes do *not* form homologous pairs (this happens in *meiosis*). Valid points are made about the second stage (which is actually *metaphase*), but the mistake about homologous chromosomes is repeated in the description of the third stage. Candidate A correctly identifies telophase as the last stage and notes that the nuclear membrane re-forms during this stage. Overall, the answer is fairly brief and lacks the detail required for a high mark. This response would be worth 5 out of 10 marks.

Overall, Candidate A scores 35 out of 70 marks for this mock paper, which would be a grade D/E.

Answers to mock paper 1: Candidate B

(1) In a water molecule, two hydrogen atoms are joined to an .....oxygen ✓.... atom by covalent bonds. When covalent bonds are formed, the .....electrons ✓.... are not always shared equally, causing a water molecule to be slightly positive at one end and slightly negative at the other. Water is therefore known as a .....dipolar ✓.... molecule. Water molecules have an attraction for each other and form ....hydrogen ✓..... bonds with neighbouring molecules.

🖉 A perfect answer, receiving 4 out of 4 marks.

(2)

Statement	Glycogen	Starch	Cellulose
Energy store in plant cells	✗	✔	✗
Glycosidic bonds present	✔	✔	✔
Polymer of α-glucose	✔	✔	✗
Unbranched chains only	✗	✔ ✗	✔

🖉 This answer receives 3 out of 4 marks. Candidate B clearly had a change of mind about whether starch consists of unbranched chains only (it does *not*, so a cross should have been inserted in this box) and has written a rather ambiguous mixture of a cross and a tick. An examiner could not tell whether it should have been a tick or a cross and so it has to be marked as wrong.

(3)

Name of organelle	Description	Function
Cell surface membrane	A phospholipid bilayer containing proteins and cholesterol ✓	To control the entry and exit of substances into/out of the cell ✓
Ribosomes ✓	Small spherical organelles, found attached to endoplasmic reticulum or free in the cytoplasm	Protein synthesis
Mitochondria ✓	Rod-shaped structures with a double membrane, the inner one folded to form cristae	Aerobic respiration ✓

🖉 An excellent answer which receives 5 out of 5 marks.

(4) (a) A = hydrophilic region ✗; B = hydrophobic region ✗

🔁 Although A *is* hydrophilic (soluble in water), the question specifically asks for the name of the *part* of the molecule, which is **a phosphate group**. B is *not* hydrophobic (it is the long fatty acid chains which are hydrophobic) and this part of the molecule is **glycerol**.

(b) The chemical bond labelled X is an ester bond ✓.

🔁 Although this answer is correct, Candidate B has wasted some time writing out the question again. It would have been enough to write 'ester bond' or even just 'ester'. The question asks you to name the bond and that is all you have to do!

(c) Unsaturated — because there are double bonds between the carbon atoms in the fatty acid chains ✓.

🔁 A very good answer, showing that Candidate B clearly understands the meaning of 'unsaturated'.

(d) glycoprotein ✓ and cholesterol ✓

🔁 Note that **carbohydrate, proteins** or **glycolipids** would also get a mark in part (d). Overall, Candidate B receives 4 out of 6 marks for this question.

(5) (a) (i) A = organic base ✓; B = ribose ✗

🔁 Part B is the pentose sugar **deoxyribose**. Ribose is found only in RNA.

(ii) X = phosphodiester bond ✓; Y = hydrogen bond ✓
(b) (i) adenine and guanine ✓
(ii) A pairs with T; C pairs with G ✓

🔁 Although this answer is biologically correct, it does not use the data in the table. Candidate B should have pointed out that **the ratio of A:T is 1:1** and **the ratio of C:G is 1:1**, suggesting that A pairs with T and that C pairs with G. Candidate B receives 5 out of 8 marks for this question.

(6) (a)

Feature	Facilitated diffusion	Osmosis	Active transport
Requires energy in the form of ATP	✗	✗	✔
Can take place against a concentration gradient	✗	✗	✔
Requires protein carrier molecules	✗✗	✗	✔

🔁 This answer receives 2 out of 3 marks. Candidate B appears to have confused facilitated diffusion with simple diffusion. Simple diffusion does *not* require protein carrier molecules, but facilitated diffusion *does* require protein carrier molecules.

(b)   Surface area   .....large...............
                 Difference in concentration .....large...............
                 Thickness of membrane   .....thin ✗...........

*This is technically correct, but the question asked Candidate B to state whether the variables would be **high** or **low** in each case and these terms should be used in the answer.*

(c)

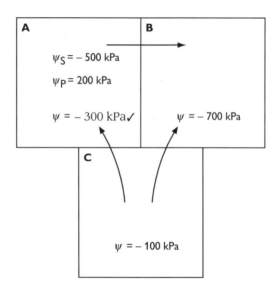

A very good answer. Candidate B has realised that you have to make two separate points for 2 marks.

(d) The uptake increases as the concentration of glucose in the surrounding fluid increases ✓; but the relationship is not linear (as expected in simple diffusion) as uptake flattens out at high concentrations of glucose ✓ because the protein carrier molecules become saturated (cannot work any faster).

An excellent answer for 2 marks, giving Candidate B 7 out of 9 marks for the whole question.

(7) (a) (i) $47 - 35 = 12$ units ✓; $(12/35) \times 100 = 34.3\%$ ✓✓

The calculations are correct, scoring Candidate B all 3 marks.

(ii) As the temperature increases from $20\,°C$ to $45\,°C$ the enzyme and substrate molecules have more energy ✓. This causes an increase in the number of collisions between enzyme and substrate molecules, leading to the formation of more enzyme–substrate complexes ✓, so the enzyme is more active ✓. Above $45\,°C$, the shape of the active site becomes distorted, the enzyme becomes denatured ✓ and enzyme activity is decreased.

A very good answer, earning Candidate B full marks for this part of the question.

(b) Repeat the experiment, but keep temperature constant using a water bath ✓. Use a range of buffers at different pH ✓, such as pH 3, pH 5, pH 7, pH 9 and pH 11 ✓. The concentration of enzyme and substrate should be kept constant ✓. The experiment should be repeated at each pH and a graph plotted of enzyme activity against pH ✓.

🖉 An excellent final answer. Overall, Candidate B scores 12 out of 12 for this question.

(8) (a) A sequence of bases in DNA ✓, which codes for the production of a particular polypeptide ✓.
   (b) CCAGUGAAGCUGGG ✓

🖉 Candidate B has correctly remembered the rules for base-pairing and that RNA does not contain thymine (T) but has uracil (U) instead. However, she has missed out one base (this answer only has 14 bases, not the 15 required). Therefore 1 mark has been lost.

   (c) proline – valine – lysine – leucine – glycine ✓✓
   (d) (i) Adenine is changed to guanine ✓ at the fifth base from the left ✓; so alanine will be coded for rather than valine.
      (ii) The amino acid sequence will be changed ✓ which will affect the primary structure of the protein ✓ and therefore change the tertiary structure and shape of the protein ✓.

🖉 A good answer, worth 3 out of 4 marks. Candidate B could have explained that changing the shape of the protein could have **changed its function**, or that the mutation may have **no effect at all**. These extra details would have given the candidate full marks. Overall, Candidate B scores 10 out of 12 marks for the whole question.

(9) Mitosis is a type of cell division concerned with growth, repair and the replacement of tissues. A cell divides to produce two new daughter cells, which are genetically identical to the parent cell. The cell divides by a series of stages: prophase, metaphase, anaphase and telophase. In prophase, the chromosomes condense and are seen to consist of a pair of chromatids joined by a centromere. Spindle fibres are produced by centrioles and the nuclear membrane breaks down at the end of prophase. The next stage is metaphase, where the chromosomes attach to spindle fibres and line up along the equator of the cell. During anaphase, the centromeres split and the chromosomes separate, moving to the poles of the cell. Telophase is the last stage, where the chromosomes reach the poles of the cell and the nuclear membrane re-forms. The cells then split and so two new cells have been produced.

🖉 Candidate B has produced a concise and detailed response to this free-prose question. This is a very good answer and would be worth 10 out of 10 marks.

Overall, Candidate B scores 59 out of 70 marks for this mock paper, which would definitely be a grade A.

ock paper

Answers to mock paper 2: Candidate A

(1)

Feature	Prokaryotic cell	Eukaryotic cell
Ribosomes	✔	✔
Endoplasmic reticulum	✔ ✗	✗ ✗
Mesosomes	✗ ✗	✔ ✗
Golgi apparatus	✔ ✗	✗ ✗

This answer receives 1 out of 4 marks. It seems that Candidate A may have known the key differences between prokaryotic and eukaryotic cells because these responses are the complete *opposite* of the correct answers. Perhaps the candidate misread the column headings, or simply confused prokaryotic and eukaryotic. Make sure that you read the question *very* carefully.

(2)

Substance	Name of test	Positive result
Protein ✓	Biuret test	Blue-black colour ✗
Reducing sugars	Emulsion test ✗	Orange-red precipitate
Starch ✓	Iodine test	Lilac colour ✗

This answer receives 2 out of 5 marks. The test for reducing sugars is the **Benedict's test** (the emulsion test is used to show the presence of *lipids*). Candidate A has also mixed up the positive results for protein and starch: protein produces a **lilac colour** in the biuret test; starch produces a **blue-black colour** in the iodine test.

(3) Lipids are fats and oils formed by ....hydrolysis ✗.............. reactions between .....glycerol ✓........ and fatty acids. During the reaction, a triglyceride is formed and three molecules of ....carbon dioxide ✗......... are removed. The main function of lipids is as an energy store, but they also play important roles in ....waterproofing ✓........ and ......insulation ✓.

This answer is worth 3 out of 5 marks. The reactions that form lipids from their constituent parts of fatty acids and glycerol are **condensation** reactions. This is true for all macromolecules — they are built up by condensation reactions and broken down by hydrolysis reactions. Three molecules of **water** are removed during this reaction.

(4) (a) Globular proteins have a tertiary structure consisting of irregularly folded polypeptide chains, giving them an approximately spherical shape. They are soluble and generally have metabolic functions, such as enzymes.

☝ This is a good description of globular proteins, but it does *not* answer the question. There is no mention of fibrous proteins and all of the stated features may also apply to them. It is much better to write *comparative* points, e.g. **globular proteins are folded into approximately spherical shapes, *but* fibrous proteins have an elongated structure**. Unfortunately, Candidate A receives no marks for this answer.

(b) Rough endoplasmic reticulum (RER) has ribosomes on its outer surface, but smooth endoplasmic reticulum (SER) does not ✓; the cisternae of RER tend to have a flattened shape and those of SER are more tubular ✓.

☝ Candidate A has made two good points for 2 marks. Note also how the answer defines a term *before* using the abbreviated version, so the examiner knows what the candidate is talking about. However, Candidate A has only made two points for a 3-mark question and should have written more, e.g. **RER is involved in protein synthesis and SER is involved in the synthesis of lipids and steroids**. Overall, Candidate A receives 2 out of 6 marks for this question.

(5) (a) They are easy to get from a plant ✗.

☝ Root tips are one part of the plant where *growth* is occurring. Therefore, there will be a lot of **cell division** taking place and the cells will show stages of mitosis.

(b) (i) in order to see the chromosomes under the microscope ✓
(ii) to break up the chromosomes ✗

☝ If you break up the chromosomes, you will not be able to see them under the microscope! This stage is to **break up the tissue** or to enable you to **see individual cells**.

(c) (i) the distance between the centromeres of the original chromosomes ✓
(ii) 10 minutes ✓

☝ This is the point when the centromeres start moving towards the poles of the cell, so curve A *decreases* to zero. At the same time, the distance *between* the centromeres of the original chromosomes starts to *increase*, as shown by line B.

(iii) anaphase is the stage in which the chromatids pull apart ✓

☝ The answer to part (c) of the question receives 3 out of 4 marks. In part (iii) no mention is made of evidence from the graph to support the answer. After 10 minutes **the distance between the centromeres and the poles decreases**, indicating that anaphase has begun. Overall, Candidate A receives 4 marks out of 7 for this question.

(6) (a) (i) the energy you have to put into a reaction ✓

☝ This is correct, though a little brief. Reference to the energy of the reacting molecules or to the minimum amount of energy required to make a reaction proceed would be needed to get both marks.

(ii)

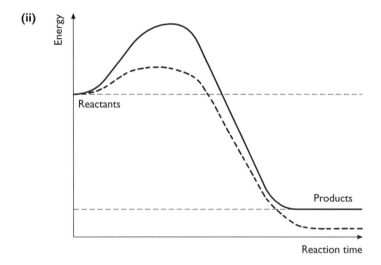

The first part of the curve is correct, as it is *below* the original. However, in this reaction the products will be at the same energy level irrespective of whether or not enzymes were used. Therefore the added curve should finish at the same level as the original, not at a lower level. This answer is worth 1 out of 2 marks.

(b) Between A and B: increasing the concentration of substrate leads to an increase in the rate of reaction ✓. Between B and C: increasing the concentration of · substrate has no effect on the rate of reaction ✓.

These responses are correct and answer the part of the question that asks you to *describe* the effect of changes in substrate concentration on the rate of reaction. However, the question also asks you to *explain* the shape of the graph, which Candidate A has not done. The increase in rate of reaction between A and B is **due to the formation of more enzyme–substrate complexes** which leads to the **formation of more product** (per unit time). Between B and C, the **enzymes are saturated and cannot work any faster**, so the **rate of reaction is limited by the concentration of enzyme** or by the **availability of active sites**. Candidate A receives 2 out of 6 marks for this part of the question, and 4 out of 10 marks overall.

(7) (a) (i) A = translation ✗; B = transcription ✗

A simple (but quite common) mistake. Candidate A has confused the terms 'transcription' (stage A: the synthesis of mRNA from DNA) and 'translation' (stage B: the synthesis of proteins on ribosomes).

(ii) RNA polymerase ✓
(iii) ribosomes ✓
(iv) mRNA is made in A and more DNA is produced in the replication of DNA ✓.

🗩 The answer to part (iv) is correct, but there are other differences between stage A (transcription) and the replication of DNA and this response would only earn 1 out of a possible 3 marks. Students often confuse transcription and DNA replication, so it is worth memorising the differences between them. Candidate A could have said: **one molecule of mRNA is produced during transcription, but two molecules of DNA are produced during replication; only one strand is copied in transcription, both strands are copied in replication; only part of DNA is copied in transcription, but the whole molecule is copied in replication**. Note that these answers emphasise the *differences* between the two processes, as asked for in the question.

(b) Guanine (G) = 33%, so cytosine (C) = 33% and C + G = 66% ✓. So adenine (A) plus thymine (T) must be 100 – 66 = 34% ✓. Therefore A = 34/2 = 17% ✓.

🗩 Candidate A has set out the calculation clearly, showing all the working. This is an excellent answer worth full marks.

(c) DNA contains deoxyribose but RNA contains ribose ✓; DNA contains the base thymine (T) but RNA contains uracil (U) instead ✓.

🗩 A clear, concise answer at the end gives Candidate A an overall score of **8 out of 12 marks** for this question.

(8) (a) The movement of molecules across a partially permeable membrane ✓

🗩 This answer is just about worth a mark, but it is a little vague. It is worthwhile learning definitions of biological terms as they are often asked for in unit tests (a biological dictionary will help). In this case, it is better to state that osmosis is the **diffusion of water molecules** across a partially permeable membrane.

(b) (i) $\psi = \psi_S + \psi_P$ ✓; i.e. $\psi = 600 + 200 = 800$ kPa ✗

🗩 Candidate A correctly remembered the formula for calculating water potential, but got the calculation wrong through forgetting the minus sign in the figure for solute potential (–600 kPa). The correct answer is therefore **–400 kPa**. Remember that water potential is *never* positive, it has a *maximum* value of zero (pure water).

(ii) Water would move out of the cell, because it moves from a higher water potential (800 kPa) to a lower water potential (–100 kPa) ✓.

🗩 Although this answer is incorrect, it still receives a mark. This is because Candidate A has based this response on the incorrect calculation in part (i) and so the conclusion is valid. It would be unfair to penalise the same mistake twice, so the mark is awarded. In fact, water would move *into* the cell because the water potential outside (–100 kPa) is higher/less negative than the water potential inside (–400 kPa).

(c) It requires energy in the form of ATP ✓. It takes place against a concentration gradient ✓.

☑ A good answer worth 2 out of 2 marks. However, although the wording of the question makes it fairly clear that 'it' in the answer refers to active transport, it would be better to write **active transport requires energy in the form of ATP but simple diffusion does not**, and so on.

(d) (i) The rate of entry of molecules into a cell is proportional to the concentration gradient, i.e. the difference between the concentration outside and that inside ✓. Therefore, up to point X, the rate of entry will increase as the external concentration increases, whether the molecules pass directly through the membrane (simple diffusion) or are transported by protein carriers (facilitated diffusion) ✓.

(ii) In cases of simple diffusion, the rate of entry of molecules into the cell will continue to increase as the external concentration increases ✓; but in facilitated diffusion the rate will slow down when the protein carrier molecules become saturated with the molecules, i.e. they cannot work any faster ✓.

☑ An excellent response to part (d), earning Candidate A 10 out of 11 marks for the whole question.

(9) Polysaccharides are made of monosaccharides. The most important polysaccharides in biology are starch, cellulose and glycogen. Starch is found as a storage material in plants and there is a lot of starch in potatoes. The test for starch is the iodine test, which gives a blue-black colour if starch is present and a yellow-brown colour if starch is not present. Cellulose is found in the cell walls of plants and acts as a food store. Glycogen is found in animal cells and a lot of it is stored in the liver. Overall, polysaccharides are very important molecules because they are used for storage and so help keep plants and animals alive.

☑ This answer is worth 4 out of 10 marks. Candidate A could have mentioned *how* polysaccharides are built up from monosaccharides and the importance of the monosaccharide subunits to the structure and properties of the polysaccharides. The answer concentrates on the storage roles of starch and glycogen, but does not say *why* they are useful for storage, i.e. that they have a compact shape, exert no osmotic effects and can be easily broken down into glucose for respiration. The reference to the iodine test is irrelevant and the statement that cellulose 'acts as a food store' is incorrect. Details about the structure of cellulose would explain *why* it has a support role in plant cell walls. Finally, reference could have been made to other polysaccharides, such as chitin in the cell walls of fungi or murein in bacterial cell walls.

Overall, Candidate A scores 37 out of 70 marks for this mock paper, which would be a grade D.

Answers to mock paper 2: Candidate B

(1)

Feature	Prokaryotic cell	Eukaryotic cell
Ribosomes	✗ ✗	✔
Endoplasmic reticulum	✗	✔
Mesosomes	✔	✗
Golgi apparatus	✗	✔

> This answer receives 3 out of 4 marks. Prokaryotic cells *do* have ribosomes, although they are slightly smaller than those found in eukaryotic cells.

(2)

Substance	Name of test	Positive result
Protein ✔	Biuret test	Blue colour ✗
Reducing sugars	Benedict's test ✔	Orange-red precipitate
Starch ✔	Iodine test	Blue-black colour ✔

> This answer receives 4 out of 5 marks. The positive result for the biuret test for proteins is a **lilac colour**. A blue colour indicates that there is *no* protein present.

(3) Lipids are fats and oils formed by ....condensation ✔.............. reactions between .....glycerol ✔........ and fatty acids. During the reaction, a triglyceride is formed and three molecules of ....water ✔......... are removed. The main function of lipids is as an energy store, but they also play important roles in ....enzymes ✗........ and ......membranes ✔.

> This answer was worth 4 out of 5 marks. Enzymes are *not* lipids, they are proteins. Some alternative correct responses to the last part of this question could be **water-proofing, protection** (of organs such as the kidney), **insulation** (to conserve heat or around nerve cells) or **steroid hormones** (such as testosterone).

(4) (a) Globular proteins have a tertiary structure consisting of irregularly folded polypeptide chains, giving them an approximately spherical shape. They are soluble and generally have metabolic functions, such as enzymes. Fibrous proteins are usually elongated and may not have a tertiary structure ✔. They are insoluble ✔ and generally have structural functions, e.g. connective tissue ✔.

> A very good answer, worth full marks.

(b) Rough endoplasmic reticulum (RER) has ribosomes on its outer surface, but smooth endoplasmic reticulum (SER) does not ✔. The cisternae of RER tend to have a flattened shape and those of SER are more tubular ✔. RER is involved in protein synthesis and SER is involved in the synthesis of lipids and steroids ✔.

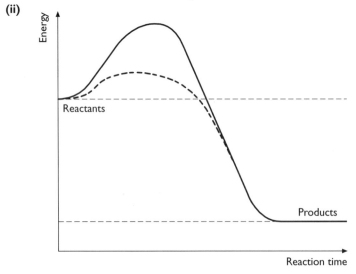 Again, a very good answer, making three separate points for 3 marks. Overall, Candidate B receives 6 out of 6 marks for this question.

(5) (a) They are growing, dividing cells and so mitosis can be seen ✓.

(b) (i) in order to see the chromosomes under the microscope ✓
(ii) to break up the tissue and so enable you to see individual cells ✓

(c) (i) the distance between the divided cells ✗

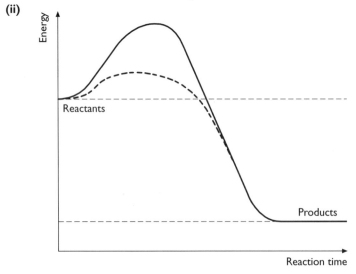 The cells have not finished dividing at the point when line B starts to increase. Therefore this line cannot represent the distance between the divided cells. It is **the distance between the centromeres of the chromosomes**.

(ii) 10 minutes ✓
(iii) Anaphase is the stage in which the chromatids pull apart and move towards the poles ✓. The graph shows that after 10 minutes the distance between the centromeres and the poles decreases, indicating that anaphase has begun ✓.

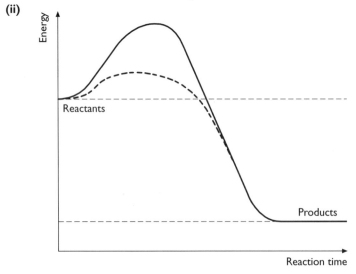 A good answer. Overall, Candidate B receives 6 marks out of 7 for this question.

(6) (a) (i) The energy required by the substrates ✓; in order that successful collisions will occur and the reaction will proceed ✓.

(ii)

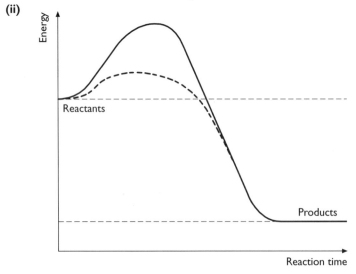

This answer is correct and receives 2 marks. The first part of the curve should be *below* the original, as enzymes *lower* the activation energy of a reaction. The products will be at the same energy level irrespective of whether or not enzymes were used. Therefore the added curve should finish at the same level as the original, as shown on the graph.

(b) Between A and B: increasing the concentration of substrate leads to an increase in the rate of reaction ✓. This is due to the formation of more enzyme–substrate complexes ✓ which leads to the formation of more product per unit time, i.e. a faster rate of reaction ✓. Between B and C: increasing the concentration of substrate has no effect on the rate of reaction ✓. This is because the temperature is limiting the rate of reaction and it cannot go any faster ✗.

🖉 This answer is worth 4 out of 6 marks. The rate of reaction levels off between B and C because the **enzymes are saturated and cannot work any faster**, so the rate of reaction is **limited by the concentration of enzyme** *or* by the **availability of active sites**. Overall, Candidate B receives 8 out of 10 marks for this question.

(7) (a) (i) A = transcription ✓; **B** = translation ✓
    (ii) polymerase ✗

🖉 Unfortunately, this response is too vague to receive a mark. There are a number of polymerases found in living organisms and the specific answer required here is **RNA polymerase**.

    (iii) ribosomes ✓
    (iv) One molecule of mRNA is produced during transcription, but two molecules of DNA are produced during replication ✓. Only one strand of DNA (the sense or coding strand) is copied in transcription, but *both* strands of DNA are copied in replication ✓. Only part of DNA (a gene) is copied in transcription, but the whole DNA molecule is copied in replication ✓.

🖉 An excellent answer, worth full marks.

(b) 18% ✗

🖉 This answer receives no marks. Candidate B might have made only one mistake but could have earned as many as 2 out of 3 marks for working. However, this answer shows no steps in the calculation. It is *vital* to show your working, as instructed in the question.

(c) DNA contains deoxyribose, but RNA contains ribose ✓. There is only one type of DNA, but three types of RNA: mRNA, tRNA and rRNA ✓.

🖉 A clear, concise answer at the end gives Candidate B an overall score of 8 out of 12 marks for this question.

(8) (a) The diffusion of water molecules ✓ across a partially permeable membrane ✓.

🖉 A concise, accurate definition worth 2 marks.

(b) (i) $\psi = \psi_S + \psi_P$ ✓; i.e. $\psi = -600 + 200 = -400$ kPa ✓
    (ii) Water would move out of the cell, because it moves from a more negative water potential (−400 kPa) to a less negative water potential (−100 kPa) ✗.

🖉 This answer is incorrect. Water would **enter** the cell, because it moves from a higher (*less* negative) water potential (−100 kPa outside the cell) to a lower (*more* negative) water potential (−400 kPa inside the cell).

(c) Active transport requires energy in the form of ATP, but simple diffusion is passive ✓. Active transport requires protein carriers to move the molecules across membranes; simple diffusion does not ✓.

e An excellent answer worth 2 out of 2 marks.

(d) (i) Diffusion and facilitated diffusion are quicker if there is a larger concentration gradient ✓. So up to a point (X), the rate of entry will increase as the concentration increases. This is true for both diffusion and facilitated diffusion ✓.

(ii) In simple diffusion, the rate at which molecules enter the cell will continue to increase as the external concentration increases ✓. In facilitated diffusion the rate slows down when the protein carrier molecules become saturated with the molecules and therefore cannot work any faster ✓.

e An excellent response to part (d), earning Candidate B 10 out of 11 marks for the whole question.

(9) Polysaccharides are made of long chains of monosaccharides, joined together by glycosidic bonds. The most important polysaccharides in biology are starch, cellulose and glycogen. Starch is found as a storage material in plants and occurs in two forms, amylose and amylopectin. Glycogen is also a storage polysaccharide, but it is found in animals. Both starch and glycogen consist of long branched chains of α-glucose. They are ideal as storage materials because they have a compact shape (so take up little room in the cell) and they have little or no osmotic effect. They can be broken down easily to provide glucose for respiration. Cellulose consists of long, unbranched chains of β-glucose and these chains are cross-linked by hydrogen bonds. This makes cellulose tough and ideal for its structural role in the cell walls of plants. Other important polysaccharides in living organisms include those found in glycoproteins in cell surface membranes, which are important in cell recognition by the immune system, and chitin found in the cell walls of fungi and the exoskeleton of insects.

e An excellent answer, worth 10 out of 10 marks.

Overall, Candidate B scores 59 out of 70 marks for this mock paper, which would be a grade A.